YOUR KNOWLEDGE HAS VALUE

- We will publish your bachelor's and master's thesis, essays and papers

- Your own eBook and book - sold worldwide in all relevant shops

- Earn money with each sale

Upload your text at www.GRIN.com and publish for free

Bibliographic information published by the German National Library:

The German National Library lists this publication in the National Bibliography; detailed bibliographic data are available on the Internet at http://dnb.dnb.de .

Imprint:

Copyright © 2018 GRIN Verlag
Print and binding: Books on Demand GmbH, Norderstedt Germany
ISBN: 9783668843950

This book at GRIN:

https://www.grin.com/document/451381

Moritz Stüber

Minute polyphagous wasps as biological pest control in European ecological fruit cultivation

GRIN Verlag

ISARA-Lyon

Agroecological Cropping Practices

Minute polyphagous wasps as biological pest control in European ecological fruit cultivation

Literature review

Career: Agroecology

Student:
Moritz Stüber

Date of submission. 23. Oct 2018

Content

Abstract

Mass releases of parasitic minute polyphagous wasps is a common biological pest control practice across the world. The application can reduce chemical insecticide use and therefore contribute to a more sustainable agriculture. Nearly all frequently used species derive from the genus *Trichogramma*. These wasps are released on several Million hectares of agricultural production, particularly in maize cultivation. This report is aiming to give an insight into a growing system, where *Trichogramma* practices are still under development. With regards to the application density, efficacy, applied species and environmental risks of *Trichogramma* in European ecological fruit cultivation was reviewed. The common application methods are lacking consistent results and pest control efficacy, since orchards differ greatly from the usual grain field cropping system. Three research directions to increase efficiency can be distinguished: new application methods (1); environmental attributes that favour *Trichogramma* (2) and suitable *Trichogramma* (mixtures) species (3). Development of equipment, suitable to disperse *Trichogramma* eggs more equally between rows is needed. A nozzle fan, spraying *Trichogramma* eggs, seemed to decrease costs and labour force by simultaneously increasing pest control consistency. Flower strips, particularly by containing buckwheat and mustard, can increase the longevity and fecundity of the wasps, leading to a better pest control performance. Local *Trichogramma* species should be preferred, as they are used to the climatic conditions and contribute to agroecological practices. The risk of mass releases on non-target insect species must be surveyed continuously. No significant effects were found, but off-field emigration does happen and needs to be monitored. The research implies the potential of *Trichogramma* to be used as an agroecological practice, being able to contribute to insecticide reduction. However, many results still need to be tested under commercial conditions.

1. Introduction

Agroecology as a science and practice is aiming to decrease the negative environmental effects of agriculture. Synthetic fertilisers and pesticides for pest control are aimed to be reduced or even completely abandoned. This brings up the question and demand for alternatives. Agroecological farming is aiming to replace agro-chemicals by biological practices, including the beneficial interactions between different organisms within the farming system (Watts and Williamson 2015). Pest control, as a concrete action, is industrially performed by spraying pesticides. Obviously, agroecological farmers need a way to protect their harvest against pests and to reduce their yield losses too. To do so, farmers can apply biological pest control, e.g. biological insecticides or predatory insects. Augmentative (mass releasing) biological pest control is utilised on more than 10°Mio. ha worldwide (van Lenteren and Bueno 2003). One of the most commonly used biological pest controls are (polyphagous) parasitic wasps. Most polyphagous wasps attack their host's eggs and initiate the larvae's death. The wasps' ability to control pests is recognised for more than 100 years, but only in the 1960's and 70's, their commercial value developed (Smith 1996). Nowadays, parasitic wasps are successfully used all around the world for inundative pest management. Parasitic wasps are categorised as natural enemies. Releasing parasitic wasps is an intervention in the relation between natural predator/prey relationships. In agricultural systems, parasitic wasps are often restoring the natural enemy balance of exotic insects, that have reached pest status in their new environment (Altieri 1995). The wasps attack eggs of different insect families but are dominantly released for the rapid reduction of Lepidoptera (particularly caterpillars) (Martin and Sauerborn 2013). The most prominent case in Europe, is the parasitisation of the European corn borer (*Ostrinia nubilalis*) in maize cultivation. Apart from mass releases to cope with massive pests, also applications in smaller scale have developed. Nowadays, minute egg wasps are e.g. often used in organic orchards (Jehle et al. 2014). The evaluation of economic success of this biological pest control measure is closely linked to its effectiveness. Nevertheless, Agroecological measures are not only evaluated by their effectiveness, but by their environmental compatibility. Literature is often giving different results on the effectiveness of parasitic wasps, depending on the type of crop and its pest, as well as on region and climate. It seems quite hard to figure out, whether the release of parasitic wasps is suitable as an effective and ecological practice. The literature review is aiming to investigate on the effectiveness of this alternative method for pest control, its application range, limitations, risks and to which extent the application is suitable as an Agroecological practice.

2. Minute polyphagous wasps as biological pest control

Minute polyphagous wasps are endoparasitoids, that develop within the eggs of their host. Polyphagous parasites do not depend on one single host but are able to vary in the choice of their host. This characteristic can be seen as the reason behind the success story of *Trichogramma*. Rearing is simplified, if the parasitized host can be selected from a wide range of species, as well as it increases the application range (Grenier et al. 2005). By consisting in approximately half a million species (most of them still unknown), the Trichogrammatidae is a diverse superfamily, of which the *Trichogramma* genus are the most commonly used wasps in agriculture. The wasps attack eggs of different insect families (butterflies, true bugs, beetles), but are mainly utilised against major moth and butterfly pests (Martin and Sauerborn 2013, Lindsey et al. 2018). To understand the whole concept of the biological pest control, the biological mechanism behind the parasitisation by *Trichogramma* will be reviewed in 2.1.

2.1. Biology

The wasp lays its eggs directly onto the host's egg, by simultaneously killing it. The next generation *Trichogramma* wasps will hatch circa ten days after. In figure 1 constitutes the biology and parasitisation mechanism of the most common example of lepidopterous pests, the European corn borer (*O. nubilalis*) and its parasite (*T. brassicae*). The moth's life cycle, as well as the parasitic interference of the *Trichogramma* wasp are shown. The (interfered) life-cycle will continue as long as host eggs are available.

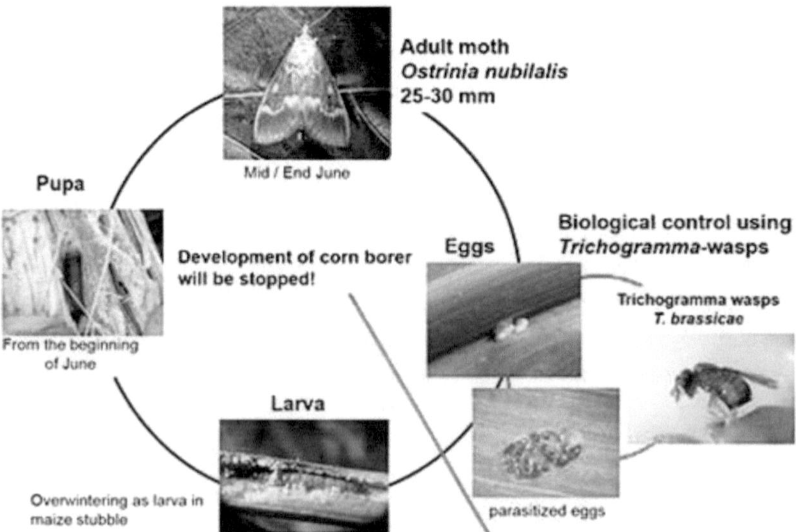

Figure 1: Life-cycle and mechanism of the *Trichogramma* wasp and its host (AMW Nützlinge GmbH 2014)

Several generations of the *Trichogramma* wasp can develop within one year. After mass releases, the half a millimetre measuring wasp usually does not survive the winter season. In nature, *Trichogramma* usually survives as a prepupae, the final larvae stadium before pupation. In the new season, increasing temperatures are necessary to complete the pupation development. When temperature rises, development is completed and the first slip of *Trichogramma* occurs. The emergence takes place temporally offset to the first warm days of the year. The annual hatching in Europe differs between mid-April and the beginning of May. In Southern regions, the emergence will happen slightly earlier (Barnay et al. 2001, Kursch-Metz 2014). *Trichogramma* in Europe is active at temperatures above 15 °C, while its optimum is between 23 and 28 °C. The activity decreases at higher temperatures. Over 32 °C, no more eggs are laid and the parasitisation activity is inhibited (AMW Nützlinge GmbH 2011, Ksentini et al. 2011).

2.2. Global perspective

Apart from the genus *Trichogramma*, only few other species are commercially used for lepidopterous pest control. Annually, *Trichogramma* wasps are applied on nearly 10 Mio. ha worldwide, being the major utilised natural enemy in biological pest control (van Lenteren and Bueno 2003, Vinson et al. 2015). On the contrary, microbial biocontrol agents such as nematodes, fungi, bacteria and viruses are estimated to be utilised on approximately 1.5 Mio. ha worldwide.

The five most common species of *Trichogramma* are: *T. evanescens, T. dendrolimi, T. pretiosum, T. brassicae,* and *T. nubilale*. In smaller scale of application and crop type, the species diversity is higher. Smith (1996) states four issues for the commercial appropriate application of *Trichogramma* species for biologic pest control. These are the "selection of the appropriate population to release, a system for mass rearing, distribution of the parasitoid, and a strategy for field release". In globally common crops, such as sugar cane, wheat, maize parasitisation rates of 60-80% can be achieved. Therefore, yield losses can be substantially reduced in a natural way. For a successful parasitisation, two applications of approximately 100,000 individuals per hectare are recommended (Wajnberg and Hassan 1994 in Martin and Sauerborn 2013). However, the right application of *Trichogramma* is depending on several factors, considering the approach (inoculative or inundative), timing and frequency. Climate, weather and season are also playing an important role. Local species should therefore be preferred, thanks to a better adaption to local climate and habitat. Using species native to their environment is the basis of the inundative theory (Smith 1996). All five most commercially relevant species are used on a global scale. *T. nubilale* and *T. pretiosum* are native to the United States, *T. dendrolimi* to China and *T. evanescens* and *T. brassicae* are native to Europe. Exotic species are used because of their commercial success and effectiveness. The use of exotic species should be limited to situations where no local species are available and a proper environmental screening is implemented prior to its application. The Agroecological idea of promoting natural enemies for pest control (Altieri 1995), is correlating with the concept of using local (natural) predators as biological pest control. The correlation is unbalanced, if exotic species are utilised. Due to exotic species and the heavy reduction of natural enemies, the predator/prey balance is often not given. In the long run, the restorage of the natural predator biodiversity is much more promising, than introducing one single predator species. Altieri (1995) is mentioning several opportunities which can lead to pest control (by parasitic wasps) in a natural and local way. Non-selective insecticides have put pressure on the parasitoids, while several studies (Hagen et al. 1971 and Stephens 1984 in Altieri 1995) proved the restorage of natural predator/prey balances, by reducing the insecticide and therefore increase natural parasitoid activity. Increasing the host diversity, either by biodiversity favouring measures (crop & vegetational diversity) or by releasing host populations, can stimulate the natural parasitoid count to increase ten-fold throughout the whole season (Parker and Pinnell 1972, Stephens 1984 in Altieri 1995). However, instead of reviewing appropriate measurse to naturally restore natural enemy biodiversity, this report will mainly focus on the agricultural practice of releasing parasitic wasps. The compatibility of *Trichogramma* species for the application in orchards and the environmental risks of mass releasing native as well as exotic species will be reviewed in the continuing.

2.3. Application in orchards

The use of polyphagous wasps in orchards is recognised for thirty years (Hassan et al. 1988). In Europe, the most prominent example of biological pest control in orchards is the protection against the Codling Moth (*Cydia pomonella*) and the Plum Moth (*Grapholita funebrana*) by *T. cacoeciae* and *T. dendrolimi* (Jehle et al. 2014). *C. pomonella* is considered as one of the main pests in the worldwide fruit growing, dedicated to several fruits, e.g. apple, peach, plum and pear cultivation (Blomefield 1989 in Samara et al. 2008). Research and commercial range of application in Central Europe has been conducted increasingly for 20 years (Zimmermann 2004). Because of upcoming resistances against common control practices, like granulovirus (CpGV) and scab-controlling Sulphur products, a diversification of pest control methods is urgently needed. The interest for commercial alternatives in organic fruit growing is increasing and drawing attention to polyphagous wasps (Kienzle et al. 2012, Jehle et al. 2014). The application of polyphagous wasps in fruit cultivation is more intense than in crop cultivation. Instead of 100,000 eggs (Wajnberg and Hassan 1994 in Martin and Sauerborn 2013) the rates of wasps released in orchards count several Millions per hectare and application. Methods depend on the size of the orchards. In small-scale orchards, cards, containing more than 1,000 *Trichogramma* pupae are attached to every or every second or third tree. The wasps are about to emerge shortly after their application. More larvae can be applied in a less developed stage, so several waves of wasps will hatch in the following weeks (Kienzle et al. 2012). The card frequency is determined by the high cost of application, since the costs per hectare and application is estimated to be around 250 to 350 € (AMW, OMYA, Rincon-Vitova, Sautter & Stepper). The research on application methods has therefore increased considerably (Zimmermann 2004). Many trials to increase the efficiency have been implemented. Generally, three different directions of efficiency increase can be pointed out: research on the application method (1), on improving the environmental attributes for *Trichogramma* (2) and on finding suitable *Trichogramma* (mixtures) species (3).

In several studies, the card release practice revealed issues in horizontal distribution. Most parasites are staying at the same tree, where the card has been attached. Here, parasitisation rates of 60-80% are obtained. Surrounding trees often have a parasitisation ratio of less than 20%, at further distance even below 5% (Wetzel 1995, Sakr et al. 2002 in Kienzle et al. 2012). Similar methods include biodegradable balls, containing the pupae, which can either be applied manually or by drone (AMW Nützlinge GmbH 2011). In recent studies, the application of bigger machinery was tested. A fan orchard sprayer, combined with nozzles, distributing the polyphagous wasps with low pressure has successfully been tested under field conditions. While the nozzles increase the distribution range, different hydrocolloid agents improve the attachment of the parasitized eggs to the trees. Best results have been achieved by a pressure of 3 to 5 bar and mixing a hydrocolloid, based on 2% Xanthan and 0,01% Tween. The fan orchard sprayer is estimated to decrease the costs per application below 100 € and also reduce manual labour force (Zimmermann et al. 2010, Kienzle et al. 2012).

The life span of adult wasps is low (Boivin 2010 in Kursch-Metz 2014). Depending on the species, the wasps live for less than a week (Hassan 1989). Studies, regarding environmental attributes mainly focus on increasing longevity and fecundity of the wasps. Planting flower strips can improve the longevity. Plants in the surrounding area, that significantly increased the longevity were buckwheat (*Fagopyrum esculentum*), mustard (*Brassica nigra*), dill (*Anethum graveolens*) and borage (*Borago officinalis*). Directly feeding honey or pollen even further increased the longevity. Buckwheat in flower strips, can also significantly increase the fecundity (up to six-fold) (Zandstra and Motooka 1978 in Romeis et al. 2005, Witting-Bissinger et al. 2008, Sigsgaard et al. 2013).

Instead of applying only one *Trichogramma* species, mixtures between different species are developed (Kienzle et al. 2012). *T. cacoeciae* and *T. dendrolimi* eggs are sometimes mixed to receive a simultaneous short- and long-term pest control. *T. dendrolimi* is often preferred, thanks to its high fecundity and a high parasitisation rate. *T. dendrolimi* has a sexual reproduction cycle, so both sexes have to be released to ensure mating before the parasitisation. Furthermore, the life span of *T. dendrolimi* is relatively low (one week) (Hassan 1989). *T. cacoeciae* is parthenogenetic, so it is possible to only release female wasps. This reduces time issues compared *T. dendrolimi* and explains the mixture of short- and long-term pest control. The parasitisation efficiency of *T. cacoeciae* is generally thought to be lower than *T. dendrolimi*. However, in some trials, single *T. cacoeciae* strains reached similar efficiencies, as the mixtures (Sakr et al. 2004). *T. evanescens* is often occurring naturally in orchards and vineyards (Barnay et al. 2001, Kienzle et al. 2012, Kursch-Metz 2014). Compared to *T. cacoeciae, T. evanescens* also seems to have a higher resistance against low and high temperatures (Scholler and Hassan 2001). The species has often not been considered in the biological pest control in orchards. *T. evanescens* is commercially offered to vineyards, controlling the Vine Moths *Lobesia botrana* and *Eupoecilia ambiguella*. In some cases, this species seemed to have a high capacity in controlling pests, especially *C. pomonella* in apple orchards (Barnay et al. 2001, Kienzle et al. 2012). In a recent study (Sigsgaard et al. 2017), *T. evanescens, T. cacoeciae* and mixtures between both species were tested on controlling *C. pomonella* pest damages. Again, *T. evanescens* proved to be most effective species, but not reducing pest damages significantly higher than the latter. The knowledge on different species and their pest control capacities is still limited (Herz et al. 2007). The species *T. aurosum*, being native to Central Europe, has revealed promising results on the control of *C. pomonella*, especially regarding temperature hardiness. Until now, the species has not been used commercially (Mansfield and Mills 2004, Samara et al. 2008, Samara et al. 2011). Several researches deal with indigenous *Trichogramma* species to eventually become commercialised, instead of studying on already established species. In many orchards, no more than a few different *Trichogramma* species were found (Herz et al. 2007, Kursch-Metz 2014), which could soon lead to a depletion of research on new capable species.

Even though, the mentioned studies focussed on native species rather than exotic ones, the environmental impact has still to be surveyed. Insect populations are decreasing across Europe. Especially pollinators, like bees or butterflies, are suffering under the use of insecticides and the decrease of their environment (European Commission 2018). Many species are already endangered and might soon be driven to extinction. Critical studies included the parasitisation on non-target lepidopterous species, as well as other species of interest (pollinators, predators). Four types of investigation can be differentiated: laboratory (1) and greenhouse assessments (2), on-field trials of non-target species parasitisation (3) and migratory movements of *Trichogramma*, possibly resulting in parasitisation off the field (4). Pollinators examined on *Trichogramma* parasitisation included many butterfly species (frequent species like: *Maniola jurtina* L. and vulnerable species like: *Polyommatus semiargus* or *Papilio machaon*) and hoverflies (*Episyrphus balteatus*). Non-target butterflies under semi-field and field conditions have a parasitisation rate ranging between 2 and 20%. The parasitisation rate of non-target species is usually two or three times lower than pest species. Especially under field conditions, the rate of non-target parasitisation is decreasing rapidly. At 2 m distance the parasitisation rate decreased below 7%, with no parasitisation occurring at 20 m distance. Lower results were found for non-lepidopterous species. Under greenhouse conditions, only 0.4% of *E. balteatus* eggs were parasitized (Babendreier et al. 2003a, Babendreier et al. 2003b). Investigation of parasitisation on predatory species was carried out for ladybirds (*Adalia bipunctata* and *Coccinella septempunctata*) and lacewings (*Chrysoperla carnea*). Parasitisation of predatory insect in semi-field and field conditions stayed below 10%. *Trichogramma* wasps were able to emerge in more than 80% of the cases, while no offspring hatched from parasitized ladybird eggs (Babendreier et al. 2003b). The effect of mass releases under actual field conditions was found to have no detrimental effects on populations of non-target species, since parasitisation is unlikely to occur. Even though, the effect of non-target parasitisation off field is low, emigration from release fields does happen (Andow et al. 1995, Follett and Duan 2012). The mobility depends on the plant species. *Trichogramma* was found to emigrate twice as often from *Trifolium pratense* than from maize, as their dedicated host plant (Babendreier et al. 2003c). Bigger migratory effects are not expected, since the wasps cover only short distance in the meter range during their lifetime (McDougall and Mills 1997). In a comparison of several natural enemies used in inundative pest control, van Lenteren et al. (2003) set up a score for risk evaluation, including establishment, dispersal, host range, environmental effects and hybridisation. *Trichogramma* is classified with the highest risk indices, due to their polyphagous and migratory characteristics, but also regarding to the increasing number of applications, especially carried out on the field by non-trained person.

3. Discussion

Trichogramma as biological pest control revealed some interesting results over the past 30 years and is applied globally in all different types of cultivation. However, the application of *Trichogramma* in Europe is only carried out on 50,000 ha. In Germany, only 1,600 ha of pome cultivation are treated with the parasitic wasps, whereas other countries are applying *Trichogramma* on several Millions of hectares. There are many reasons for the currently low number of applications. In fields of maize and other crops, plants are more equally distributed across the field than in fruit tree cultivation. Orchard trees in rows and line are separated by a few meters. Unfortunately, the difference in structure reduces the consistency of wasp application. *Trichogramma* rather moves on foot and rarely flies to new plants. The existing practices of *Trichogramma* cards or drones, are not suitable for a bigger scale fruit cultivation. In small orchards, the use of cards is still financially reasonable, but in bigger orchards, the costs per application are too high. Paired with inconsistent pest control, the application of *Trichogramma* in fruit cultivation was just not suitable for an expansion. The application of bigger machinery, like fan with integrated nozzles (Kienzle et al. 2012), needs more advocacy towards its commercialisation and to fulfil its potential of consistent efficacy by simultaneous cost reduction. In the European temperate climate zone, the activity and efficiency of *Trichogramma* is highly depending on external factors of weather and other climate conditions. Being active only above 15 °C and below 30 °C (AMW Nützlinge GmbH 2011, Ksentini et al. 2011), depending on wind and air humidity, *Trichogramma* requires a certain level of education or external instruction (van Lenteren and Bueno 2003). *Trichogramma* species are sensitive to conventional and biological insecticides, CpGV or Sulphur application for scab control. The possibility of reducing insecticides in ecological should be taken seriously, especially concerning resistance development of pests against CpGV and insecticides. A complete abandonment of insecticides seems illusive but reducing the number of sprayings by releasing *Trichogramma* is already possible. E.g. in late summer, when scab is no major problem anymore, the use of Sulphur products can be avoided and *Trichogramma* mass releases implemented (Zimmermann et al. 2010, Kienzle et al. 2012, Wang et al. 2012). The use of both, insecticides and *Trichogramma* pest control must be exactly reconciled. If fruit growers receive support and get sensitised on how to properly use the wasps as biological pest control, the application in orchards has the potential to further commercialise. Research on new *Trichogramma* species, that are suitable for the application in commercial agriculture does not seem to be promising. Species like *T. evanescens* or *T. aurosum* showed promising results, but since Central European orchards are poor in *Trichogramma* species diversity, research might quickly be depleted. The introduction of exotic species is not compatible with the idea of agroecological farming and therefore no considered option. Another issue that relates to invasiveness is the parasitisation of non-target insect species. It is of great importance not to further harm the already decreasing insect population. Studies with frequently occurring and vulnerable butterflies as well as with beneficial insects (hoverflies, ladybirds, lacewings) revealed no significant predation after mass releasing *Trichogramma*. However, *Trichogramma* mass releases should always be applied consciously and the influence on non-target insects continuously be studied. Particularly migratory effects of *Trichogramma* should stay under observation. Even though, the wasps are not likely to migrate, emigration from release fields does happen (Andow et al. 1995, Follett and Duan 2012). By improving the natural environment, e.g. by flower strips (buckwheat, mustard), natural enemies can be favoured and the predator/prey balance reconstructed. These measures favour wild insects and improve the performance of the wasps, in terms of longevity and fecundity. *Trichogramma* wasps pose an important alternative for ecological pest control in fruit cultivation, as insecticides are aimed to be reduced. If vital criteria, as using native species and monitoring the effect on the environment, are not infringed, *Trichogramma* mass releases can be considered a good practice for Agroecological fruit growing. Unfortunately, the success of this practice highly depends on future research as the performance, in terms of consistency and labour-intensity, is not yet sufficient. Particularly, the sharp decline of parasitisation in the surrounding area has to be compensated by new application methods. Otherwise, the application of *Trichogramma* in orchards will continue to stay small-scale. If *Trichogramma* application receives more attention and advocacy, new practices can be implemented and the potential of contributing to commercial ecological pest control and the reduction of insecticides can be depleted.

4. References

Altieri, M. A., 1995. Biodiversity and biocontrol: Lessons from insect pest management, in: Tommerup, I. C., Andrews, J. H. (Eds.), Advances in plant pathology. London, San Diego, 191–209.

AMW Nützlinge GmbH, 2011. Die TrichoKarte GH zur biologischen Bekämpfung von Schadschmetterlingen im Freiland und Unterglasbau, 3–9.

AMW Nützlinge GmbH, 2014. *Trichogramma* parasitic wasps. http://www.amwnuetzlinge.de/ Greenhouse-*Trichogramma*:_:44.html?language=en (retrieved October 2018).

Andow, D. A., Lane, C. P., Olson, D. M., 1995. Use of *Trichogramma* in maize: estimating environmental risks, in: Hokkanen, H. M. T., Lynch, J. M. (Eds.), Biological control: benefits and risks. Cambridge University Press, 101–118.

Babendreier, D., Kuske, S., Bigler, F., 2003a. Parasitism of non-target butterflies by *Trichogramma brassicae* Bezdenko (Hymenoptera: Trichogrammatidae) under field cage and field conditions. Biological Control 26 (2), 139–145.

Babendreier, D., Rostas, M., Hofte, M. C. J., Kuske, S., Bigler, F., 2003b. Effects of mass releases of *Trichogramma brassicae* on predatory insects in maize. Entomologia Experimentalis et Applicata 108 (2), 115–124.

Babendreier, D., Schoch, D., Kuske, S., Dorn, S., Bigler, F., 2003c. Non-target habitat exploitation by *Trichogramma brassicae* (Hym. Trichogrammatidae): What are the risks for endemic butterflies? Agricultural and Forest Entomology 5 (3), 199–208.

Barnay, O., Hommay, G., Gertz, C., Kienlen, J. C., Schubert, G., Marro, J. P., Pizzol, J., Chavigny, P., 2001. Survey of natural populations of *Trichogramma* (Hym., Trichogrammatidae) in the vineyards of Alsace (France). Journal of Applied Entomology 125 (8), 469–477.

Blomefield, T. L., 1989. Economic importance of false codling moth *Cryptophlebia leucotreta*, and codling moth *Cydia pomonella* (Linnaeus) (Lep.: Tortricidae) in an unsprayed apple orchard in south Africa. African Entomology 5, 319–336.

Boivin, G., 2010. Reproduction and Immature Development of Egg Parasitoid, in: Consoli, F. L., Parra, J. R. P., Zucchi, R. A. (Eds.), Egg Parasitoids in Agroecosystems with Emphasis on *Trichogramma*. Springer Science+Business Media B.V, Dordrecht, 1–23.

European Commission, 2018. The EU approach to tackle pollinator decline. http://ec.europa.eu /environment/nature/conservation/species/pollinators/index_en.htm (retrieved October 2018).

Follett, P. A., Duan, J. J., 2012. Nontarget Effects of Biological Control: Springer US, 57–163.

Grenier, C., Gomes, S. M., Febvay, G., Bolland, F., 2005. Artificial diet for rearing Trichogramma wasps (Hymenoptera: Trichogrammatidae) with emphasis on protein utilization. Second International Symposium on Biological Control of Arthropods, 480–486.

Hagen, K. S., van den Bosch, R., Dahlsten, D. L., 1971. The Importance of Naturally-Occurring Biological Control in the Western United States, in: Huffaker, C. B. (Eds.). Biological Control, Springer US, Boston, MA, 253–293.

Hassan, S. A., 1989. Selection of suitable *Trichogramma* strains to control the codling moth *Cydia pomonella* and the two summer fruit tortrix moths *Adoxophyes orana, Pandemis heparana* [Lep.: Tortricidae]. Entomophaga 34 (1), 19–27.

Hassan, S. A., Kohler, E., Rost, W. M., 1988. Mass production and utilization of *Trichogramma*: Control of the codling moth *Cydia pomonella* and the summer fruit tortrix moth *Adoxophyes orana* (Lep.: Tortricidae). Gesunde Pflanzen (Germany, F.R.), 413–420.

Herz, A., Hassan, S. A., Hegazi, E., Khafagi, W. E., Nasr, F. N., Youssef, A. I., Agamy, E., Blibech, I., Ksentini, I., Ksantini, M., Jardak, T., Bento, A., Pereira, J. A., Torres, L., Souliotis, C., Moschos, T., Milonas, P., 2007. Egg parasitoids of the genus *Trichogramma* (Hymenoptera, Trichogrammatidae) in olive groves of the Mediterranean region. Biological Control 40 (1), 48–56.

Jehle, J. A., Herz, A., Keller, B., Kleespies, R. G., Koch, E., Larem, A., Schmitt, A., Stephan, D., 2014. Statusbericht Biologischer Pflanzenschutz 2013, 75–103.

Kienzle, J., Zimmermann, O., Wührer, B., Triloff, P., Morhard, J., Landsgesell, E., Zebitz, C. P. W., 2012. New species and new methods of application: a new chance for *Trichogramma* in Codling Moth Control? 15th Intl. Conf. on Organic Fruit-Growing, 317–321.

Ksentini, I., Herz, A., Ksantini, M., Jardak, T., Hassan, S. A., 2011. Temperature and strain effects on reproduction and survival of *Trichogramma oleae* and *Trichogramma cacoeciae* (Hymenoptera: Trichogrammatidae). Biocontrol Science and Technology 21 (8), 903–916.

Kursch-Metz, T., 2014. Untersuchungen zur Agrobiodiversität von nützlichen Insekten am Beispiel von *Trichogramma*-Schlupfwespen, 35–40.

Lindsey, A. R. I., Kelkar, Y. D., Wu, X., Sun, D., Martinson, E. O., Yan, Z., Rugman-Jones, P. F., Hughes, D. S. T., Murali, S. C., Qu, J., Dugan, S., Lee, S. L., Chao, H., Dinh, H., Han, Y., Doddapaneni, H. V., Worley, K. C., Muzny, D. M., Ye, G., Gibbs, R. A., Richards, S., Yi, S. V., Stouthamer, R., Werren, J. H., 2018. Comparative genomics of the miniature wasp and pest control agent *Trichogramma pretiosum*. BMC biology 16 (1), 54.

Mansfield, S., Mills, N. J., 2004. A comparison of methodologies for the assessment of host preference of the gregarious egg parasitoid *Trichogramma platneri*. Biological Control 29 (3), 332–340.

Martin, K., Sauerborn, J., 2013. Agroecology: Springer Netherlands, 236–237.

Parker, F. D., Pinnell, R. E., 1972. Further Studies of the Biological Control of *Pieris rapae*: Using Supplemental Host and Parasite Releases. Environmental Entomology 1 (2), 150–157.

Romeis, J., Babendreier, D., Wäckers, F. L., Shanower, T. G., 2005. Habitat and plant specificity of *Trichogramma* egg parasitoids — underlying mechanisms and implications. Basic and Applied Ecology 6 (3), 215–236.

Sakr, H. E. A., Hassan, S. A., Zebitz, C. P. W., 2002. Control of the Codling moth *Cydia pomonella* by releasing mass reared egg parasitoids of the genus *Trichogramma*. Mitt BBA (389), 26–35.

Sakr, H. E. A., Hassan, S. A., Zebitz, C. P. W., 2004. A new baiting device to capture and monitor the activity of *Trichogramma* in the field. Egyptian Journal of Biological Pest Control 14 (1), 37–42.

Samara, R., Monje, J. C., Zebitz, C. P. W., Qubbaj, T., 2011. Comparative biology and life tables of *Trichogramma aurosum* on *Cydia pomonella* at constant temperatures. Phytoparasitica 39 (2), 109–119.

Samara, R. Y., Carlos Monje, J., Zebitz, C. P.W., 2008. Comparison of different European strains of *Trichogramma aurosum* (Hymenoptera: Trichogrammatidae) using fertility life tables. Biocontrol Science and Technology 18 (1), 75–86.

Scholler, M., Hassan, S. A., 2001. Comparative biology and life tables of *Trichogramma evanescens* and *T. cacoeciae* with *Ephestia elutella* as host at four constant temperatures. Entomologia Experimentalis et Applicata 98 (1), 35–40.

Sigsgaard, L., Betzer, C., Naulin, C., Eilenberg, J., Enkegaard, A., Kristensen, K., 2013. The effect of floral resources on parasitoid and host longevity: Prospects for conservation biological control in strawberries. Journal of insect science (Online) 13, 104.

Sigsgaard, L., Herz, A., Korsgaard, M., Wührer, B., 2017. Mass Release of *Trichogramma evanescens* and *T. cacoeciae* Can Reduce Damage by the Apple Codling Moth *Cydia pomonella* in Organic Orchards under Pheromone Disruption. Insects 8 (2), 2-9.

Smith, S. M., 1996. Biological Control with *Trichogramma*: Advances, Successes, and Potential of Their Use. Annual Review of Entomology 41 (1), 375–406.

Stephens, C. S., 1984. Ecological upset and recuperation of natural control of insect pests in some Costa Rican banana plantations. Turrialba (IICA) 34, 101–105.

van Lenteren, J. C., Babendreier, D., Bigler, F., Burgio, G., Hokkanen, H. M. T., Kuske, S., Loomans, A. J. M., Menzler-Hokkanen, I., van Rijn, P. C. J., Thomas, M. B., Tommasini, M. G., Zeng, Q.-Q., 2003. Environmental risk assessment of exotic natural enemies used in inundative biological control. BioControl 48 (1), 3–38.

van Lenteren, J. C., Bueno, V. H.P., 2003. Augmentative biological control of arthropods in Latin America. BioControl 48 (2), 123–139.

Vinson, S. B., Greenberg, S. M., Liu, T., Rao, A., Volosciuk, L. F., 2015. Biological control of pests using *Trichogramma*: Current status and perspectives. 1st ed. Xianyang: Xi bei nong lin ke ji da xue chu ban she, 496.

Wajnberg, E., Hassan, S. A., 1994. Biological control with egg parasitoids. CAB International, Wallingford, UK, 304.

Wang, Y., Yu, R., Zhao, X., Chen, L., Wu, C., Cang, T., Wang, Q., 2012. Susceptibility of adult *Trichogramma nubilale* (Hymenoptera: Trichogrammatidae) to selected insecticides with different modes of action. Crop Protection 34, 76–82.

Watts, M., Williamson, S., 2015. Replacing chemicals with biology: Phasing out highly hazardous pesticides with agroecology. Penang, Malaysia: Pesticide Action Network Asia and the Pacific, 2-5.

Wetzel, C., 1995. Untersuchungen zum Einsatz von *Trichogramma dendrolimi* Matsumura (Hym., Trichogrammatidae) zur Bekämpfung von Tortriciden im Apfelanbau. Berlin, Wien: Blackwell-Wiss.-Verl, 29–67.

Witting-Bissinger, B. E., Orr, D. B., Linker, H. M., 2008. Effects of floral resources on fitness of the parasitoids *Trichogramma exiguum* (Hymenoptera: Trichogrammatidae) and *Cotesia congregata* (Hymenoptera: Braconidae). Biological Control 47 (2), 180–186.

Zandstra, B. H., Motooka, P. S., 1978. Beneficial effects of weeds in pest management—a review. Pest Articles and News Summaries, Section A, 24, 333–338.

Zimmermann, O., 2004. Use of *Trichogramma* wasps in Germany: Present status of research and commercial application of egg parasitoids against lepidopterous pests for crop and storage protection. Gesunde Pflanzen (Germany, F.R.) 56 (6), 157–166.

Zimmermann, O., Wührer, B., Kienzle, J., Triloff, P., Zebitz, C. P. W., 2010. Utilization of *Trichogramma*-wraps to control the codling moth *Cydia pomonella* with a spraying application release technique. 57. Deutsche Pflanzenschutztagung "Gesunde Pflanze - gesunder Mensch", Humboldt-Universität zu Berlin, 217.

YOUR KNOWLEDGE HAS VALUE

- We will publish your bachelor's and
  master's thesis, essays and papers

- Your own eBook and book -
  sold worldwide in all relevant shops

- Earn money with each sale

Upload your text at www.GRIN.com
and publish for free